WORLD BANK TECHNICAL PAPER NUMBER 3

Ventilated Improved Pit Latrines: Recent Developments in Zimbabwe

Peter R. Morgan and D. Duncan Mara

TECHNOLOGY ADVISORY GROUP WORKING PAPER—Number Two

Technology Advisory Group
The World Bank
Washington, D.C. 20433

This is a document published informally by the World Bank. In order that the information contained in it can be presented with the least possible delay, the typescript has not been prepared in accordance with the procedures appropriate to formal printed texts, and the World Bank accepts no responsibility for errors. The publication is supplied at a token charge to defray part of the cost of manufacture and distribution.

The views and interpretations in this document are those of the author(s) and should not be attributed to the World Bank, to its affiliated organizations, or to any individual acting on their behalf. Any maps used have been prepared solely for the convenience of the readers; the denominations used and the boundaries shown do not imply, on the part of the World Bank and its affiliates, any judgment on the legal status of any territory or any endorsement or acceptance of such boundaries.

The full range of the World Bank publications is described in the *Catalog of World Bank Publications*. The *Catalog* is updated annually; the most recent edition is available without charge from the Publications Distribution Unit of the Bank in Washington or from the European Office of the Bank, 66, avenue d'Iéna, 75116 Paris, France.

First Printing: December 1982.

Morgan, Peter R.
   Ventilated improved pit latrines.
   (TAG working paper ; WP/02)
   Cover title.
   Bibliography: p.
   1. Privies.   I. Mara, D. Duncan (David Duncan),
1944-   II. Technology Advisory Group.
III. Title.   IV. Series.
TD775.M67 1982   628'.744   82-15935
ISBN 0-8213-0078-4

Abstract

This paper describes recent developments in Zimbabwe in the design of ventilated improved pit (VIP) latrines. Two basic designs are presented, one suitable for peri-urban areas and the other, a low-cost version of the first, for rural areas. The peri-urban VIP latrine consists of a circular pit (1.5 m diameter, 3 m deep) fully or partially lined with cement mortar and with, at its top, a brick collar on which is supported a 1.9 m diameter, 75 mm thick concrete slab precast on site. A 1.8 m high spiral shaped superstructure is constructed on the cover slab in ferrocement or brick and a flat roof slab placed on top. A 150 mm diameter asbestos cement or polyvinyl chloride vent pipe with a fly screen at its top is then erected immediately adjacent to the outside of the superstructure (alternatively a brick vent pipe may be used in conjunction with a brick superstructure). The total cost, including labor and materials, ranges between US$150 and $160, depending on the superstructure and vent pipe materials. A commercial kit version of this design is also described. The rural VIP latrine consists of a rectangular pit (1.5 m x 0.6 m x 3 m) over which are placed longitudinal and transverse wooden logs of around 100 mm diameter which are then covered with anthill soil and a thin layer of cement mortar. A spiral superstructure is then built in mud and wattle, thatch, soil or local bricks and covered with a conically shaped thatched roof. The vent pipe is made from local reeds, fitted with a fly screen and rendered with cement mortar. The cost of the rural VIP latrine (excluding the cost of freely available traditional building materials) is US$8. Both the peri-urban and rural VIP latrine designs have been found to be socially acceptable in Zimbabwe (where some 20,000 have been built) and very effective in eliminating odors and controlling fly breeding in the latrine.

TABLE OF CONTENTS

PREFACE

In 1976 the World Bank commenced a two-year research program into appropriate water supply and sanitation technologies suitable for implementation in low-income urban and rural communites in developing countries. The objective of undertaking such a program was that the Bank and other international and bilateral agencies might be better informed on alternative, low-cost technologies so that their investments in water supply and sanitation would be better able to benefit the very large number of low-income communities whose immediate need for these basic services is so great. The results of this research program have been published as the initial twelve reports in the series entitled "Appropriate Technology for Water Supply and Sanitation"; these reports are listed in Annex IV to this document.

Following this Bank research program, the United Nations Development Programme, in preparation for the International Drinking Water Supply and Sanitation Decade (1981-1990), initiated Global Project GLO/78/006 in November 1978, with the World Bank as executing agency in order to translate these research results into actual projects. The objectives of this Global Project were to assist governments in developing water supply and sanitation projects which were responsive to the needs of low-income urban fringe and rural areas, which the beneficiaries could afford, which maximized public health benefits, and which could be realized and widely replicated within institutional, financial and socio-cultural constraints. The Project also helped governments identify suitable sources of funds for implementation. In January 1982 the Global Project was succeeded by Interregional Project INT/81/047 with essentially the same objectives. The Project is currently active in a number of developing countries in Africa, Asia and South America and the project team – the Technology Advisory Group (TAG) – is multi-disciplinary, comprising sanitary engineers, tropical public health specialists and social scientists (with particular expertise in cultural anthropology and health education).[1]/

While the Bank was undertaking this research program, it was aware of the execellent work in pit latrine design being done at the Blair Research Laboratory in Zimbabwe (then Southern Rhodesia). At that time it was not possible for the Bank to evaluate the work being done there. However a TAG mission went to Zimbabwe in April 1981 and this report is the result of the cooperation achieved then between Dr. Peter Morgan, who pioneered the development of the ventilated improved pit latrine in Zimbabwe, and Dr. Duncan Mara, Professor of Civil Engineering at the University of Leeds and TAG's Technical Adviser.

---

[1]/ Further information on the Interregional Project and TAG's activities may be obtained from the Project Manager, UNDP INT/81/047, Transportation and Water Department, the World Bank, 1818 H Street, N.W., Washington, D.C. 20433, United States of America.

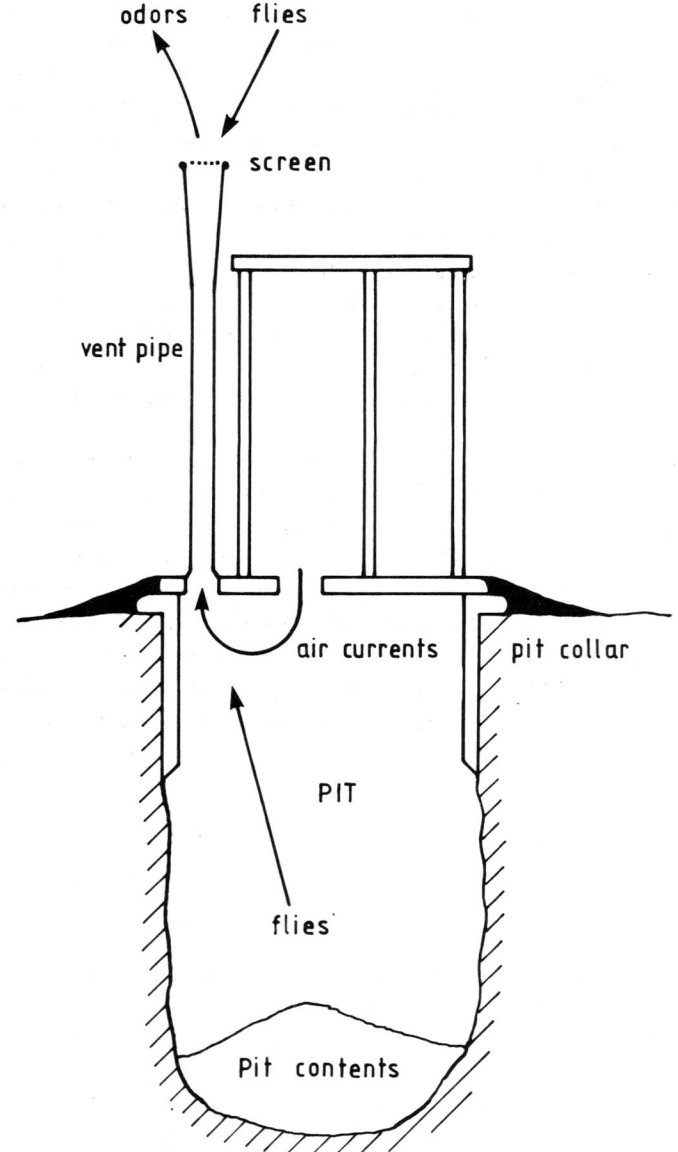

Figure 1: Schematic diagram of a ventilated improved pit latrine.

INTRODUCTION

1.   Excreta-related diseases are responsible for a large proportion of the morbidity and mortality in developing countries, especially amongst low-income communities in urban fringe and rural areas where adequate water supplies and sanitation facilities are typically absent. Excreta control is thus of paramount importance if the incidence of these diseases is to be reduced. Research sponsored by the World Bank[1] has clearly shown that excreta-related diseases can be greatly reduced by (a) the provision of sanitary toilets, of whatever type, which people of all ages will use and keep clean; (b) the effective treatment of excreta or sewage prior to discharge or reuse; (c) the provision of an adequate water supply such that water consumption is in the region of 30 to 50 liters per capita per day, which is normally the minimum requirement for the control of those excreta-related infections which have a water-washed mode of transmission; and (d) an effective and sustained program of personal hygiene education by the responsible local authority.

2.   Economic and financial constraints dictate that the water supply and sanitation technologies to be used for the control of excreta-related diseases in low-income communities must be affordable by these communities; these technologies must therefore have low capital and operating costs. In rural areas and in urban areas up to a population density of around 300 persons per hectare, the least cost technically feasible sanitation technology will often be the ventilated improved pit (VIP) latrine, and there is no doubt that the VIP latrine will be one of the sanitation technologies most widely adopted during the International Drinking Water Supply and Sanitation Decade to meet the needs of the urban and rural poor.

VENTILATED IMPROVED PIT LATRINES

General Description

3.   Traditionally-designed pit latrines have two main disadvantages: they smell and give rise to serious fly nuisance. Both these disadvantages are substantially reduced in VIP latrines. As shown schematically in Figure 1, the pit of the VIP latrine is slightly offset from the superstructure in order to permit the installation of a vertical screened vent pipe. As explained below, both fly and odor nuisance are controlled by the vent pipe; in all other respects VIP latrines are similar to, and designed in the same way as, traditional pit latrines, although some recent designs have the novel feature that the pit is emptyable so that the latrine can be a permanent structure[2].

1/   See Annex IV, Volume 3.

2/   See TAG Working Paper on Ventilated Improved Pit Latrine Design (forthcoming).

Odor Control

4.   There are two explanations of the vent pipe's role in odor control: (a) the thermal effect of solar radiation on the pipe's external surface and (b) the suction effect of wind across the top of the pipe. The relative importance of these two ventilation mechanisms is currently unknown, although field investigations are at present being conducted with TAG assistance in three developing countries. In due course the results of these studies will be published in this series.

5.   **Solar Radiation.** The effect of solar radiation is to heat up the vent pipe and thus the air inside it. This air becomes less dense and therefore rises up out of the vent pipe, and is replaced by cooler air from below. In this way a strong circulation of air is created through the superstructure and pit and thence up the vent pipe. Any odors emanating from the fecal material in the pit are thus drawn up the vent pipe, so leaving the superstructure odor-free.

6.   **Wind.** The effect of wind passing across the top of the vent pipe is to create a negative (suction) pressure within the pipe, so that air is drawn out and replaced by air from below, thus establishing the air circulation pattern described above.

7.   It is apparent that both ventilation mechanisms may operate at the same time, although clearly the solar radiation effect can only occur during daylight hours. In spite of the present incomplete understanding of how the vent pipe actually works (and thus how the vent pipe can be optimally designed), the latrines developed in Zimbabwe, which are described below, have performed very well, with odors being completely eliminated.

Insect Control

8.   **Flies.** Flies are attracted to pit latrines by the odors emanating from them. In VIP latrines flies are attracted to the top of the vent pipe since that is where the odors come from. If the vent pipe is covered with a fly screen, the flies are unable to enter and lay their eggs. However a few flies will enter the pit via the superstructure and eventually new adult flies will emerge from the pit. Newly emergent flies are phototropic and thus, provided the superstructure is reasonably dark, they will fly up the vent pipe since the only light they can see is that at the top of the vent pipe. They are prevented from leaving, however, by the fly screen and in time they fall back into the pit and die. Early experiments in Zimbabwe [1] showed that this form of fly control is very effective: in a 78 day monitoring period, 13 953 flies were caught from an unvented pit latrine, but only 146 were caught from a vented (but otherwise identical) pit latrine.

---

[1]   P.R. Morgan (1976). The pit latrine - revived. **Central African Journal of Medicine, 23, 1-4.**

9.   **Mosquitoes.** Wet pits encourage mosquito breeding, although in Zimbabwe this is not generally a severe problem. The ventilation system of the VIP latrine reduces mosquito breeding but not to the extent that fly breeding is reduced. Covering the surface water in wet pits with polystyrene balls has been found to be an effective mosquito control strategy 1/. This work has been recently confirmed in Zimbabwe, where 1 kg of 4-6 mm diameter polystyrene balls added to wet pits of 1.76 m^2 cross sectional area achieved substantial mosquito control 2/; however the long term efficacy and practicality of this method of mosquito control and its effect on sludge accumulation rates in pits subject to seasonally variable groundwater levels remain to be determined. Recent work in Tanzania 3/ suggests that mosquito control can also be achieved by placing a suitably designed trap over the squatting plate hole; such a strategy is necessary since mosquitoes are not so phototropic as flies and so may emerge through the squat hole, especially in the evenings 4/. Further research is underway on mosquito control in wet pits.

10.   The vent pipe thus performs three vital functions: it eliminates odors in the superstructure, prevents most flies from entering the pit and traps newly emergent adults. It is important that air circulation through the latrine is not impeded in any way, for example by placing a cover over the squat hole. Such covers used to be recommended to control flies, but in VIP latrines they are not only unnecessary but also detrimental and their use should be discouraged 5/.

1/   P. Reiter (1978). Expanded polystyrene balls: an idea for mosquito control. **Annals of Tropical Medicine and Parasitology,** 72(6), 595-596.

2/   Experimental results are given in Annex III.

3/   C.F. Curtis (1981). Insect traps for pit latrines. **Mosquito News, 40**(4), 626-628.

4/   Recent work in Botswana and Tanzania has shown that approximately two-thirds of emerging mosquitoes try to leave via the vent pipe and one-third leave via the squat hole (C.F. Curtis and P.M. Hawkins, "Entomological studies of on-site sanitation systems in Tanzania and Botswana," Transactions of the Royal Society of Tropical Medicine and Hygiene, **76**(1), 99-108; 1982).

5/   An exception to this rule may be in areas where it is culturally unacceptable to have a dark superstructure interior and therefore a squat hole cover is needed to reduce the amount of light entering the pit from the superstructure. Research is required to compare the effect on fly control of having no cover (thus maximizing air flow but permitting light to enter through the squat hole) as against having a cover (thus impeding air flow but restricting the light which is needed to encourage young flies to try to exit up the vent pipe). If a cover is used, it should be raised from the slab so that air circulation is not unduly inhibited.

Figure 2: Early VIP latrine design with a door.

ZIMBABWEAN VIP LATRINE DESIGNS

Latrine Entrance

11.   The first VIP latrines built in Zimbabwe in the mid-1970s were designed with a wooden door (Figure 2). This was found to have several disadvantages: wood is expensive, the hinges rust and often the door is left open with the result that the superstructure is not kept dark and consequently flies emerge via the squat hole, rather than being trapped in the vent pipe. There have also been instances where the door has been removed and chopped up for firewood.

12.   The design of the superstructure was later modified to a spiral shape (Figures 3 and 4), so that a door is no longer necessary. The superstructure thus always remains dark and consequently fly control is continuously effective. There are four different spiral designs currently in use in Zimbabwe: ferrocement and brick versions, a mass produced ferrocement kit version and a low-cost version which may be made from mud and wattle, thatch or low-cost bricks; the low-cost version is especially suitable for low-income communities in rural areas. All these designs have been found to be socially acceptable in rural areas of Zimbabwe, where some 20 000 VIP latrines have been built. Privacy is ensured by the practice, which developed spontaneously, of knocking on the superstructure wall before entering; a knock given in reply indicates that the latrine is in use.

Ferrocement Spirals

13.   The ferrocement spiral VIP latrine (Figure 3) comprises (a) a partially-lined circular pit, normally dug to a depth of 3 m; (b) a brick collar; (c) a 75 mm thick concrete cover slab which has two holes, one for the vent pipe and the other as a squat hole; (d) a spiral ferrocement superstructure; (e) a roof slab; and (f) a mass produced vent pipe of either asbestos cement or unplasticized polyvinyl chloride (uPVC) formulated so as to be stable to ultra-violet radiation. These components, together with constructional details, are described below. Working drawings and a schedule of materials are given in Annexes I and II respectively.

14.   **The Pit.** For family latrines the pit is dug to a depth of 3 m and with a diameter of 1.5 m; the diameter is increased to 1.75 m or more for communal units used in schools, prisons etc. At the top of the pit a brick "ring beam" is made by laying a single circular course of bricks in cement mortar (5 parts builder's or river sand 1/, 1 part cement). In very firm soils which do not flood during the rainy season, the pit wall can be adequately lined to a depth of 1 m by plastering a 10 mm layer of cement mortar directly on to the soil face. In less stable soils or in high

---

1/   In southern Africa sand is usually described as either pit (or quarry) sand or river sand to indicate its origin. Pit sand has a high proportion of very fine material, with generally much more than 3% passing a British Standard No. 200 sieve (0.074 mm). Builder's sand refers to pit sand that has been graded to remove most of the fine material so that it closely resembles river sand in its particle size distribution.

Figure 3: Ferrocement spiral VIP latrine with asbestos cement vent pipe.

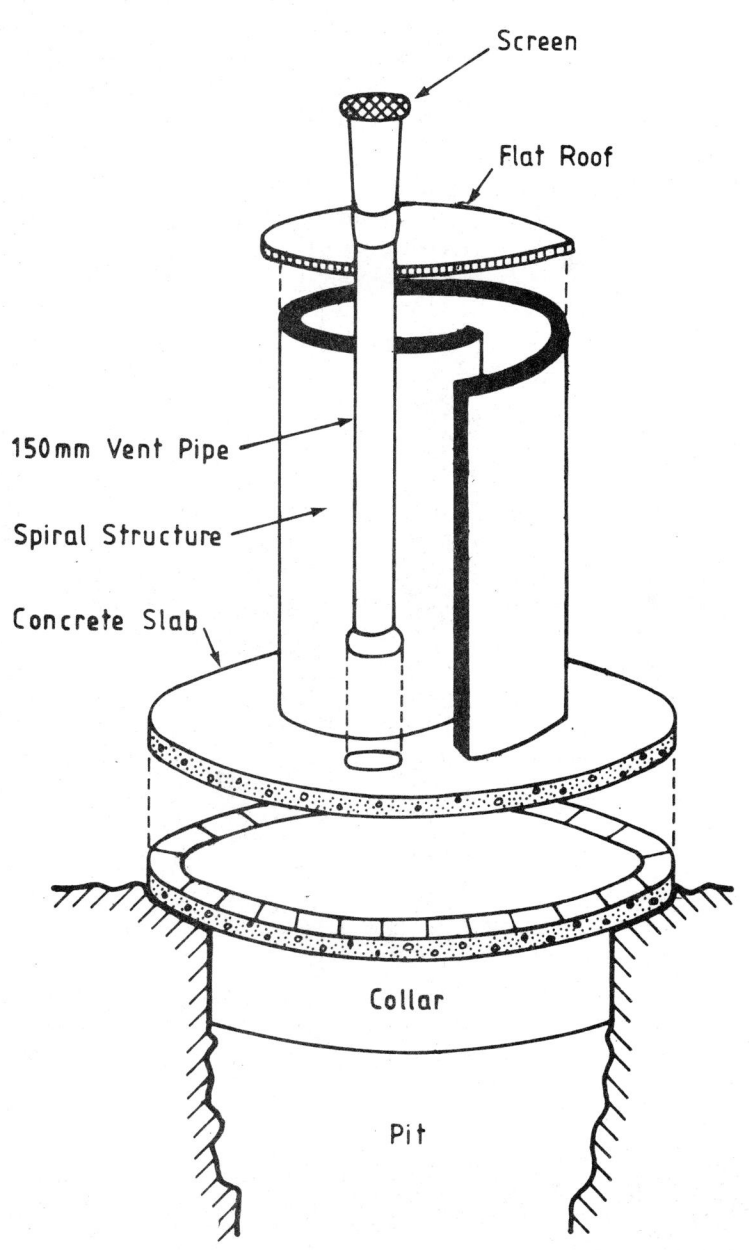

Figure 4: Exploded schematic diagram of ferrocement spiral VIP latrine.

ground water table areas it is necessary to extend this lining to the base of the pit. The mixture to cover the collar and first meter is 5 parts sand to 1 part cement. The lining below the 1 m level can be plastered with an 8:1 mix.

15.   In Zimbabwe pits were initially designed with a capacity of 0.087 m^3 (3 ft^3) per person per year. This is now realized to have been much too conservative. In family latrines examined in Zimbabwe, sludge accumulation rates rarely exceed 0.02 m^3 per person per year where the latrine is regularly washed down and paper of some sort is used for anal cleansing. Figures much lower than this have been recorded, but there is much variation depending on whether the facility is used as a washroom and how much refuse material is thrown down the pit. In Zimbabwe VIP latrines are often used as washrooms because they do not smell. Calculations for sludge build-up are based on the total number of users including children. This figure rises when the pit is always dry and when solid objects are used for anal cleansing; under these conditions accumulation rates can be doubled. Thus, where soil conditions permit and where there is no risk of polluting shallow aquifers used for potable water supplies, it is desirable that the latrine is also used as a washroom.

16.   Assuming that a family latrine is used until the pit is full to within 0.5 m below ground level, a 1.5 m diameter pit, 3 m deep should last a family of 6 for approximately 35 years if it is regularly washed down or used as a washroom. Dry pits of a similar size should last for about 20 years. School latrines which remain wet have a very extensive life since the sludge digests without the addition of wastes during periods of school holidays[1].

17.   **Cover Slab.** A circular concrete slab, 75 mm thick and 0.4 m larger in diameter than the pit, is precast on site. Two apertures are left in the slab for the squat hole and the vent pipe, as shown in Annex I. A plywood or steel mold is useful if large numbers of slabs are required. Alternatively, for small numbers, the slab may be cast on a plastic sheet within a circle of bricks; templates made from 100 mm strips of sheet metal are used for the two apertures.

18.   The concrete is made from 4 parts 18 mm aggregate, 2 parts builder's or river sand and 1 part cement; a water:cement ratio of approximately 0.4 should be used. The concrete is placed in the mold to mid-depth; a grid of 6 mm mild steel reinforcing bars at 150 mm centres is then laid and the remaining concrete added and trowelled flat. The slab is cured for a minimum of 3 days and then placed over the pit; it is bedded on to the brick ring beam with cement mortar to ensure an airtight fit. The correct orientation of the cover slab is important, and this is discussed in paragraph 25 below.

---

[1]   It is because solids accumulation rates in Zimbabwean VIP latrines are so low that no work has been done on designing desludgeable pits. In periurban areas it can be assumed that after some 20 years the latrine would be upgraded (see Annex IV, Volume II) and this would obviate the need for desludging. In rural areas the mud and wattle VIP latrine (see paragraphs 33 et seqq.) can be readily dismantled and erected again over a new pit; this option will normally be considerably cheaper than desludging.

19.   **Superstructure.** Formwork for the spiral ferrocement superstructure is available commercially in Zimbabwe; it is made from 1.8 m wide corrugated galvanized iron sheeting and 40 mm steel angle bars. Initially the formwork was designed in three parts, but now four parts are used (Annex I). The formwork is located in the appropriate position on the cover slab and covered with 38 mm mesh chicken wire, which is secured to the formwork at top and bottom with 8 swg (3.12 mm) wire. Cement mortar (5 parts builder's sand[1]/, 1 part cement) is then plastered on the formwork to just cover the mesh; after 1-2 hours a second layer of mortar is plastered on, to give a total thickness of 40-50 mm. After two days in moderately warm weather, the formwork can be removed. The ferrocement, once it has dried, is cured for about a week by regularly soaking it with water. Once the formwork has been removed, the cover slab is plastered with cement mortar so that a step is formed at the entrance (to keep out rainwater) and the slab surface is sloped in all directions towards the squat hole. If required, foot rests can be added at this stage; current practice in Zimbabwe, however, is not to provide foot rests as they are not demanded by the users.

20.   **Roof.** The roof is made to the same shape as the superstructure but 50 mm larger all round. The construction procedure is in general the same as for the cover slab, although the slab is thinner (25 mm) and made from cement mortar (3:1 mix) reinforced with a single layer of chicken mesh. When the roof slab is cured, it is bedded in on top of the superstructure with cement mortar.

21.   **Vent Pipe.** Early trials in Zimbabwe showed that 100 mm diameter asbestos cement vent pipes did not perform well, even when provided with a top section enlarged to 150 mm. However a 150 mm diameter pipe enlarged at its top to 200 mm was found satisfactory and this has been adopted as the standard design. The enlarged top section was incorporated into the design to compensate for the reduction in effective cross-sectional area (and thus greater air flow head loss) due to the flyscreen. Two standard vent pipes are commercially available in Zimbabwe, one made of asbestos cement, the other of ultra violet stabilized PVC. Originally pipes were made of galvanized iron, but these tend to corrode after use for a number of years. The asbestos pipe comes in two sections: a lower section, 1.9 m x 146 mm internal diameter with a standard collar, and an upper expansion section, 0.74 m long tapering from 146 mm to 216 mm internal diameter (see Figures 2 and 3 and Annex I). The two sections are cemented together using stiff cement mortar (2 parts sand, 1 part cement). A glass fiber flyscreen is glued with epoxy resin to the top of the pipe, and held in place by an asbestos cement ring. The PVC pipe is 2.44 m long and the main shaft has an internal diameter of 155 mm and an external diameter of 162 mm. The external diameter of the coned section is 200 mm. The pipes are sold already black-pigmented. As with the asbestos cement pipe a glass fiber screen is glued with a PVC ring to the head of the pipe. There is no recess to cause the accumulation of leaves and other debris. When the flyscreen has been fitted, the pipe is fitted vertically

---

1/   Alternatively equal parts of river sand and pit sand may be used. In Zimbabwe it has been found that river sand by itself does not adhere easily to the mold, and pit sand alone does not permit the mortar to develop sufficient strength.

over the ventilation hole in the cover slab and mortared in position; 3 mm diameter galvanized wire is used to tie the vent pipe to the top of the superstructure.

22.   Glass fiber flyscreens are used in Zimbabwe as they have been found to be more durable than aluminum screens and less expensive than brass or stainless steel.   In Zimbabwe glass fiber screens have lasted for at least seven years without requiring replacement.   The flyscreen material most commonly used is imported from Australia[1]/; the fiber diameter is 0.342 mm and it has approximately 550 fibers per meter length in one direction and 675 in the other (aperture size: approximately 1.5 x 1.2 mm).

23.   **Brick Spiral Latrines.** VIP latrines have been successfully built in Zimbabwe with brick superstructures.   In place of the ferrocement spiral, twenty courses of 20 bricks are laid in a spiral shape [2]/; a plywood template (Annex I) is useful to indicate the position of the first course.   The interior and exterior may be rendered with cement mortar, if required; in Zimbabwe usually only the interior is rendered.   In all other respects, the latrine is similar to the ferrocement version described above, although the vent pipe can be made in brickwork as well: twenty five courses of 6 bricks are laid in the form of a chimney to leave an internal cross section of 225 mm x 225 mm.   Brick vent pipes have the advantage that they retain heat longer than either PVC or asbestos cement pipes and can thus maintain a thermally induced circulation of air well into the night.   Recently a VIP latrine has been developed which uses only bricks and mortar.   The cover slab is made in the form of an arch, with formwork made from local reeds; details are given in Annex I.   The roof slab is made as described in paragraph 20.   This type of latrine, shown in Figure 5, requires approximately 900 bricks.

24.   **Painting.** To protect the cement-rendered cover slab, it is painted with a fairly thick coat of black bitumastic paint; the interior walls are also painted to a height of around 50 cm.   To keep the superstructure dark, the ceiling is also painted black, and in order to maximize the absorption of solar radiation so is the external surface of the asbestos cement vent pipe.

25.   **Latrine Orientation.** It is very important that the opening of the spiral superstructure should not face either east or west when the morning or evening sunlight can penetrate the interior of the superstructure and so provide emergent flies with an alternative source of light.   Whether the opening should face north or south is generally decided by the relative positions of the house and the latrine, in order to provide maximum privacy. The vent pipe should ideally face the equator so that it receives the most solar energy.

26.   The latrine should be located at least 2 m from trees or overhanging branches, as these interfere with the proper operation of the vent pipe.

---

[1]/   Cyclone K-M Products Pty. Ltd., Wire Cloth Division, Melbourne, Australia.

[2]/   If the bricks are laid on edge, only 15 courses of 20 bricks are required.

Figure 5: VIP latrine with brick superstructure, brick vent pipe and brick arch cover slab. (Note: the design has now been altered so that the superstructure wall forms one side of the vent pipe ; see Annex I, drawing 5)

Figure 6: Cobwebs at the top of a vent pipe.

27. **Maintenance.** ·The cover slab needs to be cleaned regularly, and occasionally it needs to be repainted with bitumastic paint. The area around the latrine should be kept free from vegetation (especially climbing plants which may stop solar heating of the vent pipe). The flyscreen should be inspected periodically to check that it is still intact. In Zimbabwe leaves and other items rarely settle on the flyscreen to restrict the air flow. Cobwebs have been observed in pipes that have been left lying on the ground prior to construction; these should be washed through with water. Cobwebs are often found around the top outside rim of the vent pipe but rarely, if ever, occlude the pipe itself or the mesh. Spiders learn quickly that flies are attracted to the top of the vent pipe (Figure 6); lizards have also learnt that this is a good place to wait for food.

28. **Costs.** Current materials costs in Zimbabwe are indicated in the schedule of materials given in Annex II; these amount to Z$ 63 (US$ 96) for the ferrocement version; Z$ 76 (US$ 116) for the brick design with a mass produced AC or PVC vent pipe; and Z$ 69 (US$ 105) for the brick design with a brick vent pipe[1]/. The cost of the ferrocement version assumes that the superstructure formwork can be used for 50 latrines. Depending on the local soil conditions, between two and nine man-days of unskilled labor are required for excavation of a 1.5 m diameter x 3 m pit and three and six man-days of skilled and unskilled labor respectively for the cover slab and complete superstructure (see Annex II). At current Zimbabwean rates (Z$ 130 and 30 per month for skilled and unskilled labor respectively), the total labor costs per latrine are Z$ 33 (US$ 53) (assuming two man-days for excavation). Total costs are thus as follows:

Ferrocement VIP latrine with AC or PVC vent pipe	Z$ 99	US$ 150
Brick VIP latrine with AC or PVC vent pipe	Z$ 109	US$ 166
Brick VIP latrine with brick vent pipe	Z$ 102	US$ 155

29. **High Groundwater Areas.** In some parts of Zimbabwe, the groundwater table is close to the surface at certain times of the year. In these areas it has been found that raising the cover slab 0.5 m above ground level is an effective strategy (Figure 7). Under such conditions the vent pipe has been found to be still able to control odors and flies, even when the water table is very close to ground level.

30. **Current Design Trends.** Consideration is presently being given to (a) the design of a moveable spiral superstructure for use in low-density urban areas where there is space on each plot for at least two alternating pit sites; (b) a cover slab which facilitates the upgrading of the latrine to a pour-flush toilet[2]/; and (c) improving the efficiency of digestion in pits by

---

1/ Zimbabwean costs have been converted at the April 1981 rate of exchange: Z$ 1 = US$ 1.52.

2/ See Annex IV, Volume 2.

Figure 7: VIP latrine with raised superstructure for use in high groundwater table areas.

lining the pit walls and base with cement mortar and connecting the pit to a soakaway with a short length of pipe (initial observations in Zimbabwe of this modified VIP latrine suggest that it could be very suitable for periurban areas where housing densities are high).

31.   **Commercial Latrine Kit.** A version of the ferrocement spiral VIP latrine is commercially available 1/ in Zimbabwe (Figures 8 and 9). The kit comprises (a) a preshaped superstructure spiral sheet of 100 mm steel mesh (bar diameter: 4 mm) tightly covered in cotton or hessian fabric; (b) a 250 mm vent pipe also made from 100 mm steel mesh and similarly covered; (c) a 1.5 m diameter circular roof sheet of the same material and covering; and (d) formwork with integral reinforcing for the cover slab (1.5 m diameter) and with apertures for the vent pipe and the squat hole. The pit is normally excavated to a depth of 3.3 m and a diameter of 1.1 m. The pit is lined to at least 1 m with cement mortar and a brick ring beam is laid as described in paragraph 14. When the kit is delivered to the site concrete is placed in the cover slab mold; when the slab has cured it is placed over the pit and the superstructure and roof steelwork then placed in position. Cement mortar (1 part cement, 2 parts sand and a proprietary additive to increase workability; 0.5-0.6 water/cement ratio) is applied by brush to both sides of the superstructure and roof fabric in thin layers to give a total thickness between 20 and 25 mm. The vent pipe is similarly coated on the outside and, when dry, placed in position; the whole latrine superstructure is then given a final application of cement slurry.

32.   The commercial latrine kit costs Z$ 60 (US$ 91) 2/. Labor costs add Z$ 27 (US$ 41), assuming 2 man-days for pit excavation; materials (four bags of cement, sand, aggregate and bricks) add a further Z$ 21 (US$ 32). Thus the total cost of the commercial kit latrine is Z$ 108 (US$ 164).

Rural Spirals

33.   Although the spiral latrines described above work extremely well in practice, their costs are too high for subsistence farmers in the rural areas of Zimbabwe, and therefore four very low-cost VIP latrines have been recently developed. They are all based on local house building skills and require, apart from traditional rural housing components, only a 50 kg bag of cement, a fly screen, nails, tying wire and 0.5 liter of black bitumastic paint; these items cost a total of Z$ 5 (US$ 8).

34.   **Mud and Wattle Spiral Latrine.** This latrine, shown in Figures 10 and 11, comprises (a) a rectangular pit; (b) a wooden cover slab; (c) a mud and wattle spiral superstructure; (d) a thatch roof; and (e) a cement rendered vent pipe made from reeds. A working drawing and schedule of materials are given in Annexes I and II respectively.

1/   Kitform Shelters and Sanitation (Pvt.) Ltd., PO Box AY.51, Harare.

2/   April 1981 prices; kit price f.o.r. Harare and exclusive of local sales tax (currently 10%).

Figure 8: VIP latrine made from commercially available latrine kit.

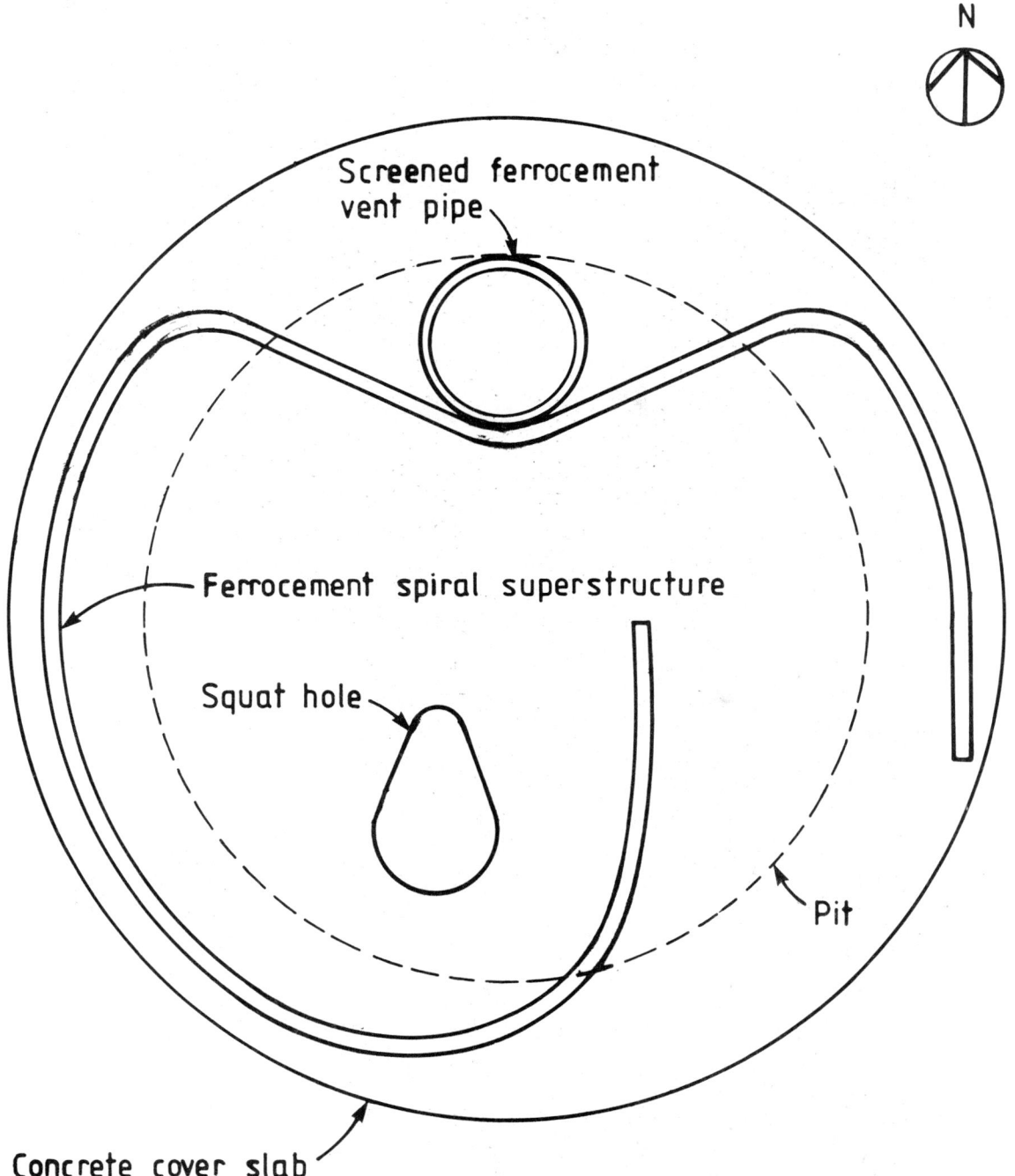

Figure 9: Sketch plan of the commercial VIP latrine shown in Figure 8.

Figure 10: Mud and wattle spiral VIP latrine.

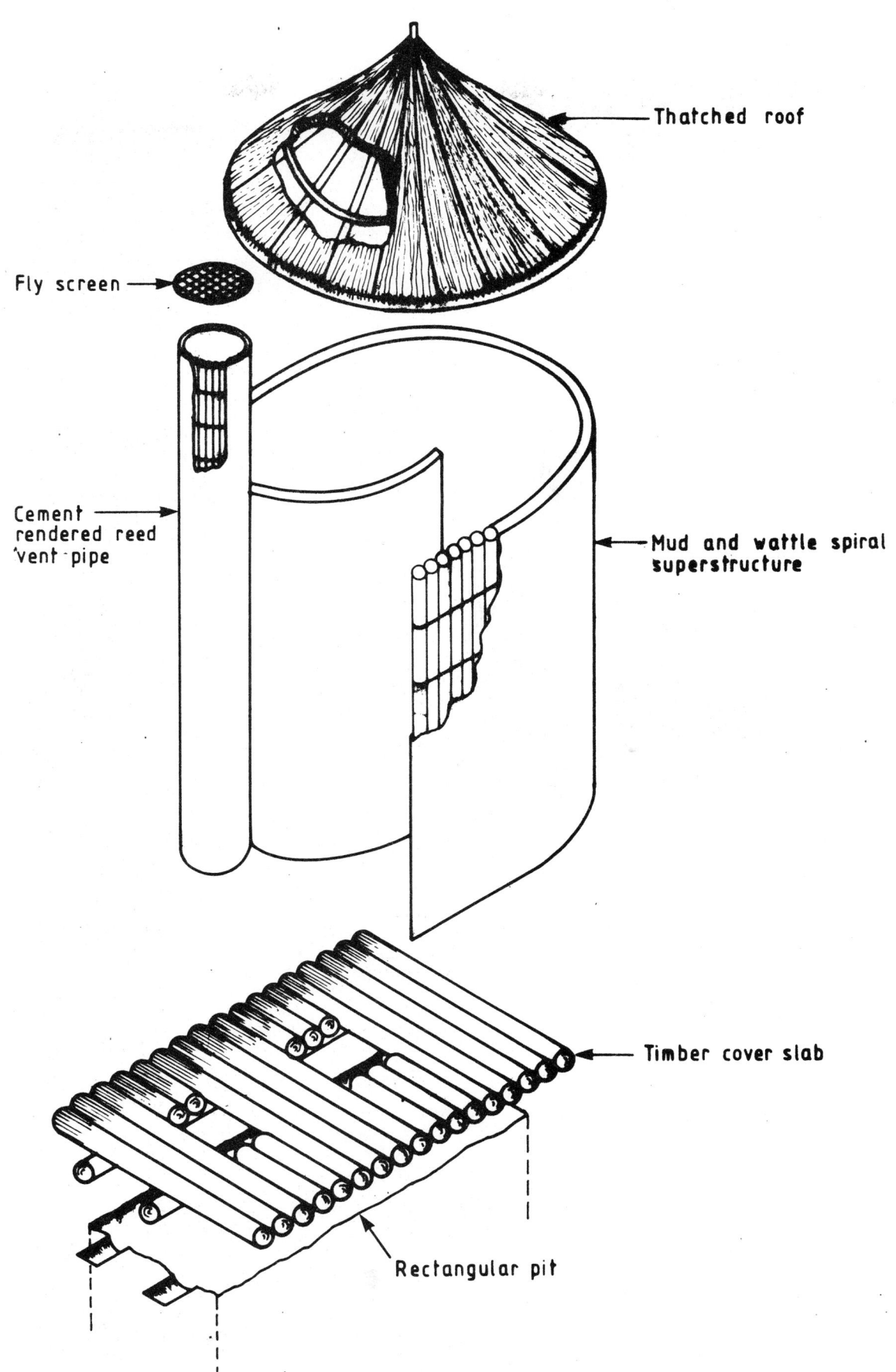

Figure 11: Exploded schematic diagram of mud and wattle spiral VIP latrine.

35.   The pit dimensions are 1.5 m x 0.6 m x 3 m. It is important to ensure that the longitudinal axis of the pit lies north-south, to permit correct orientation of the opening (see paragraph 25). Once the pit has been excavated (and, if necessary, lined 1/), the cover slab is formed. This is done by placing two logs, measuring 2.1 to 2.3 m long and approximately 100 mm in diameter, along the pit 300 mm apart, such that their upper surface is flush with ground level (this necessitates removing soil at both ends of the pit). Logs measuring 1.2 m long and roughly 100 mm in diameter are then placed across the longitudinal logs without gaps and nailed or tied to them; apertures for the vent pipe and squat hole are formed at the appropriate places (see Annex I) by using pairs of shorter logs which come to the inner edge of the longitudinal logs. The wooden logs used should be resistant to termite and fungal attack; in Zimbabwe mopane (Colophospermum mopane) and mususu (Terminalia sericea) are commonly used 2/.

36.   Once the logs are in position, the superstructure is then built. Some 30 to 40 timber poles, 1.8 m long and 50 to 80 mm in diameter, are erected in the spiral shape, nailed to the cover slab and tied together using 18 swg (1.219 mm) wire. The lower ends of some of the poles should be roughly cut to a point so that they may be firmly wedged between and nailed to the cover slab logs. The upper sections of the poles are kept in place by fastening rings of green saplings around them. The roof is then made from gum poles about 30 mm in diameter which are pliable and can be easily shaped to the desired circular form. The diameter of the roof base is 2 m and its apex 0.5 m above the plane of the base. The roof is made by weaving and tying 1.2 m long gum poles between five circles of green saplings 225 mm apart. The roof is then thatched with straw or dry grass and placed on and tied to the superstructure. This procedure was adopted as it is the traditional method for making roofs in rural Zimbabwe. The thatching has to be very dense to keep the superstructure sufficiently dark for good fly control.

37.   Once the superstructure and roof is complete the application of mud begins; in Zimbabwe traditional practice in the rural areas is not to use soil from the ground but from termite hills as this is found to have better adhesive properties and greater durability. The superstructure is first plastered with mud, both inside and outside. The cover slab is then also plastered with mud such that the floor slopes in all directions to the squat hole. As the mud dries, cracks appear and the surfaces are plastered with mud again to fill these cracks and to provide increased strength. The mud is allowed to dry out and all surfaces are then plastered with a thin coat of

---

1/   Pit lining (see paragraph 14) requires an additional half bag of cement (Z$ 1.7, US$ 2.5).

2/   In rural areas of developing countries local knowledge of suitable timbers and termite protection methods is generally very good, and it is always worth asking the local people what timbers they use and where they use them in buildings (see P.A. Campbell, "Some developments in tropical timber technology", Appropriate Technology, 2 (3), 21-23, 1975). In Zimbabwe less resistant woods such as the indigenous Msasa (Brachystegia spiciformis) or gum wood (Eucalyptus spp.) are commonly protected against termite attack by coating them with liberal quantities of wood ash, used engine oil, coarse salt, carbolinium or dieldrin.

cement mortar (1 part cement, 6 parts sand). The cover slab is then painted with black bitumastic paint.

38.   The vent pipe is constructed from a 2.4 m x 0.9 m (8 ft x 3 ft) mat of local reeds woven with string or wire. The mat is rolled up around four or five 280 mm diameter rings of green saplings to form a vent pipe of approximately 28 cm internal diameter (Figure 12), and the fly screen is wired on to one end. The vent pipe is then plastered around half its circumference with cement mortar; when this has dried it is placed in position and tied to the superstructure, and then the rest of the vent pipe is plastered.

39.   Finally, the exposed parts of the cover slab are covered with soil which is placed so as to slope gradually away from the latrine to the surrounding ground level. Grass is then planted to provide protection against the rain.

40.   **Thatched Latrine.** This latrine, shown in Figure 13, is very similar to the mud and wattle latrine, the only difference being in the superstructure. The spiral is made from gum poles placed at approximately 150 mm centres and held in position with horizontal saplings, also at 150 mm centres, which are interwoven with and tied to the vertical members. The exterior is densely thatched to exclude light. This version of the VIP latrine is especially suitable in areas where timber is in short supply.

41.   **Anthill Latrine.** In areas where grass and poles are very scarce, the superstructure can be made of well-kneaded anthill soil built up in the form of sausages to the spiral shape. The vent pipe is made in a similar fashion, coils of anthill soil being wound in a circle to form the tube.

42.   **Low-Cost Brick Latrine.** Locally made burnt bricks are commonly available in rural areas of Zimbabwe and their cost is a quarter of that of factory-made bricks. They can be satisfactorily used to build a spiral superstructure over the rectangular pit; the cover slab and thatch roof are made as described above in paragraphs 35 and 36.

43.   **Maintenance.** The rural spiral VIP latrines require regular maintenance to the cover slab and superstructure; this involves repairing any wear and tear to the slab, walls, roof and vent pipe. Since the architectural style of these latrines is essentially the same as that of their houses, the householders have the necessary skills to do regular maintenance work on the latrines; normally this is done once a year after the rainy season. The only maintenance work about which instruction is needed is the periodic inspection and replacement, if necessary, of the fly screen.

Training and Education

44.   Several methods are currently being used in Zimbabwe to extend knowledge and public awareness of the VIP latrine. These include the following:

    (1) A description is included in the school curriculum (grade 6). Models are built in classrooms. In an ongoing program, schools in the rural areas are being serviced with VIP latrines.

Figure 12: Reed mat rolled up to form low-cost vent pipe.

Figure 13: Thatched spiral VIP latrine.

(2) Two films - one on the ferrocement VIP latrine, the other on the mud and wattle design - have been made for the Home Services Mobile Cinema unit, which has a target audience of 1.5 million viewers per year in the rural areas. The films are in English and the two most common local languages, Shona and Ndebele.

(3) Instruction leaflets are available in English, Shona and Ndebele.

(4) Health Assistants of the Ministry of Health are trained in VIP latrine construction techniques.

(5) Many demonstration VIP latrines have been built throughout the country. The design is used by many government ministries at their field stations.

(6) Training courses in VIP latrine construction are held at the Henderson Research Centre, near Harare, where all the original experimental work on VIP latrines was carried out.

Design Transfer

45.   The transfer of these VIP latrine designs to other countries requires sociocultural care. For example in some societies the direction of the spiral, that is whether the spiral is dextral or sinistral[1]/, may be an essential consideration at the design stage. Other societies may not like the spiral shape and prefer a more "linear" design; this can be easily accommodated, as can the inclusion of a door which may be mandatory in certain cultures. The material used for the superstructure is not particularly important, provided the interior can be kept sufficiently dark for good fly control.

46.   The substructure design described in paragraph 14 - lining the pit walls with cement mortar - has been found perfectly satisfactory in most parts of Zimbabwe. This is due to the very high positive cohesive and good frictional properties of the soils, which are for the most part residual soils derived from igneous rocks (mainly granite, gabbro, epidyrite, gneiss and basalt). Only in Matabeleland, which borders Botswana and thus the Kalahari Desert, are the soil conditions such that pit lining in honeycomb brickwork is necessary. Substructure design in other countries must, of course, take into account local soil conditions.

47.   In rural areas it is best to design the latrine as far as possible in the same way as the local houses are constructed, so that self-help construction and maintenance can be used with only the minimum of external instruction and supervision. Such an approach is not only likely to be the least cost one, but it also ensures that the latrines blend in well with their

---

1/   A dextral spiral latrine has its opening on the right of the squat hole when viewed from in front of the opening (i.e., one enters the latrine in an anti-clockwise direction); a sinistral latrine has it on the left, with entrance in a clockwise direction.

environment; such aesthetic consideration may well prove to be one of the more important factors affecting local acceptance and sustained use of latrines in rural areas.

WORKING DRAWINGS

This Annex contains eight working drawings. Drawing ZVIP/01 shows the general arrangement of the ferrocement and brick spiral designs; drawing ZVIP/02 gives the cover slab and pit lining details for both north-opening and south-opening versions of these designs; and drawing ZVIP/03 details the spiral geometry and roof slab for these latrines and also the plywood template used to mark out the first course of the brick design. The superstructure mold for the ferrocement spiral VIP latrine is shown in drawing ZVIP/04, and details of the brick arch design (which obviates the need for a concrete cover slab) are given in drawing ZVIP/05. The general arrangement of the mud and wattle rural VIP latrine is shown in drawing ZVIP/06, and the ZVIP/07 drawing gives the cover slab and pit details for the rural latrine. The final drawing ZVIP/08 details the asbestos cement, PVC, brick and rendered reed vent pipes.

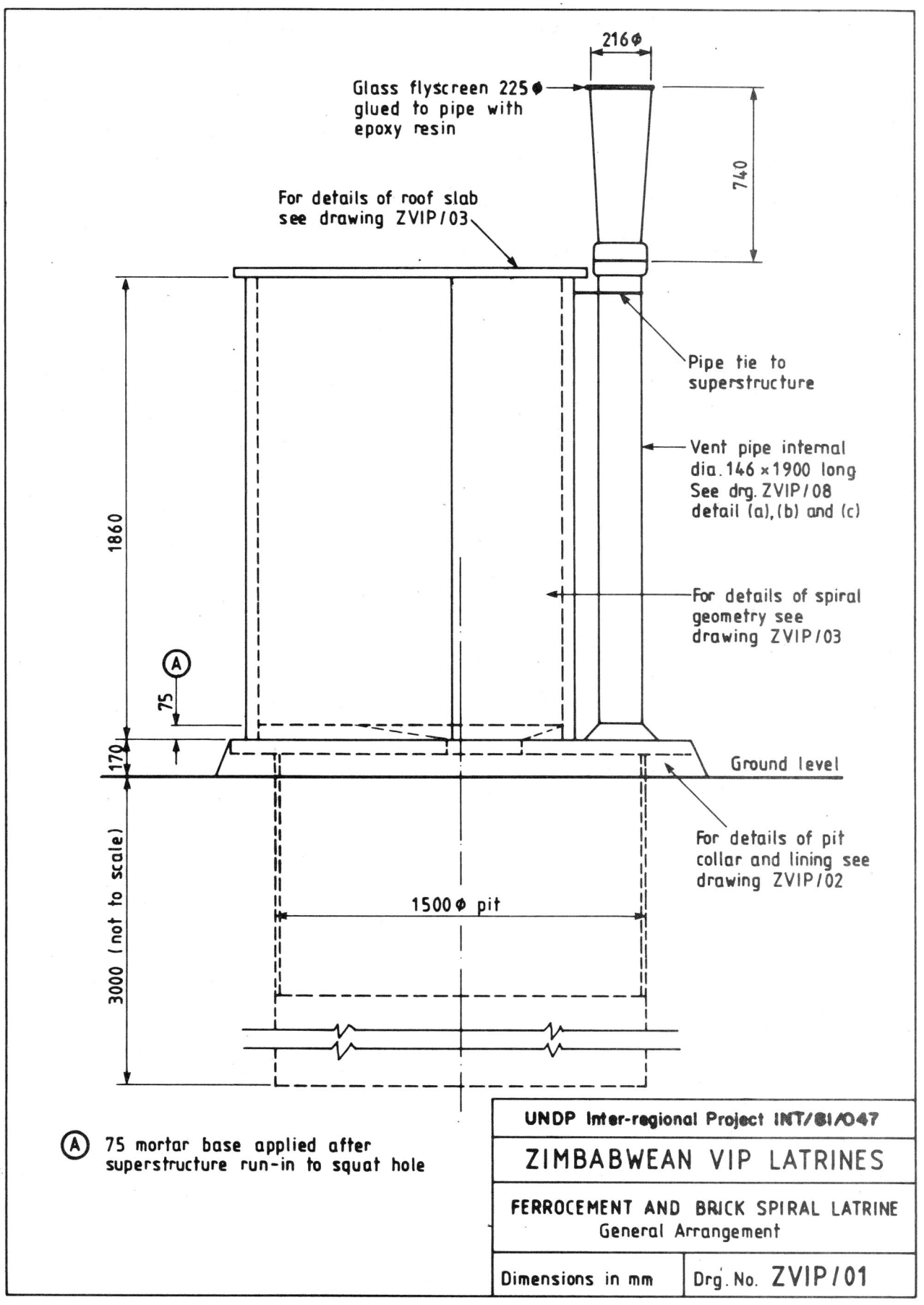

216⌀

Glass flyscreen 225⌀
glued to pipe with
epoxy resin

740

For details of roof slab
see drawing ZVIP/03

Pipe tie to
superstructure

Vent pipe internal
dia.146 ×1900 long
See drg. ZVIP/08
detail (a), (b) and (c)

1860

For details of spiral
geometry see
drawing ZVIP/03

Ⓐ

75

170

Ground level

3000 (not to scale)

For details of pit
collar and lining see
drawing ZVIP/02

1500 ⌀ pit

Ⓐ 75 mortar base applied after
superstructure run-in to squat hole

UNDP Inter-regional Project INT/81/047
ZIMBABWEAN VIP LATRINES
FERROCEMENT AND BRICK SPIRAL LATRINE General Arrangement
Dimensions in mm · Drg. No. ZVIP/01

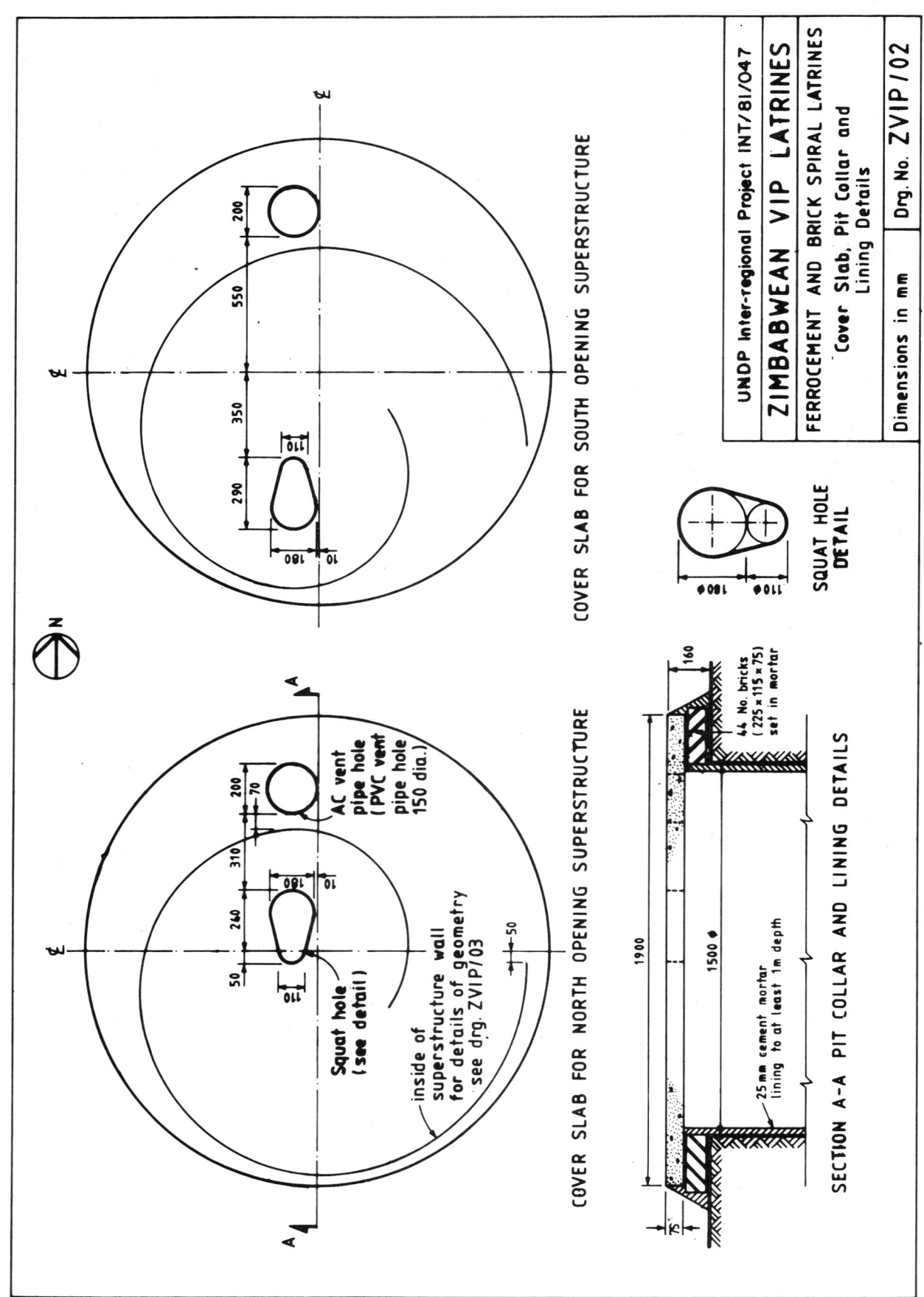

COVER SLAB FOR SOUTH OPENING SUPERSTURCTURE

SQUAT HOLE DETAIL

COVER SLAB FOR NORTH OPENING SUPERSTRUCTURE

AC vent pipe hole (PVC vent pipe hole 150 dia.)

Squat hole (see detail)

inside of superstructure wall for details of geometry see drg. ZVIP/03

44 No. bricks (225 × 115 × 75) set in mortar

25 mm cement mortar lining to at least 1m depth

SECTION A-A PIT COLLAR AND LINING DETAILS

UNDP Inter-regional Project INT/81/047

ZIMBABWEAN VIP LATRINES

FERROCEMENT AND BRICK SPIRAL LATRINES
Cover Slab, Pit Collar and Lining Details

Dimensions in mm   Drg. No. ZVIP/02

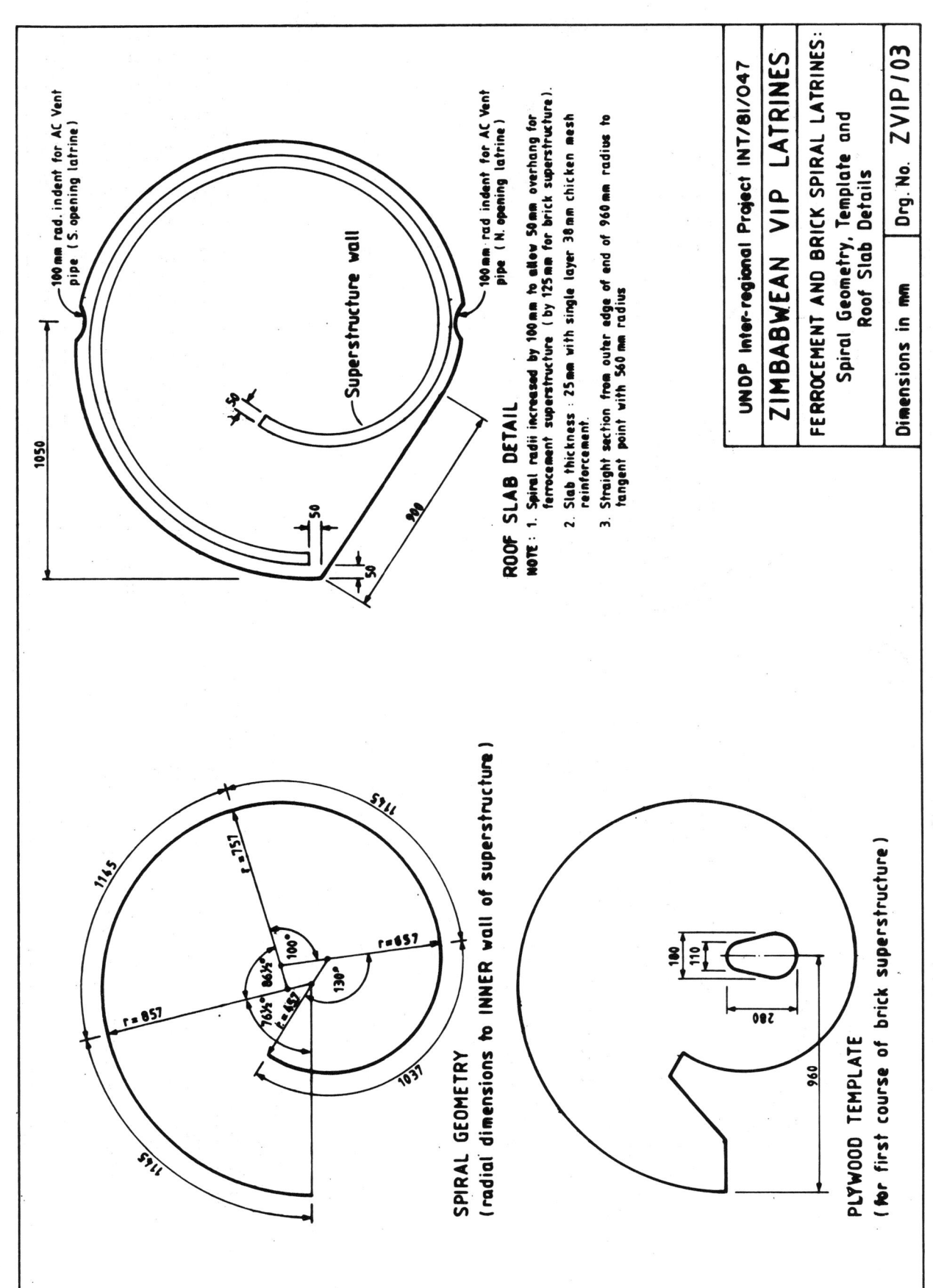

ROOF SLAB DETAIL

NOTE : 1. Spiral radii increased by 100mm to allow 50mm overhang for
ferrocement superstructure (by 125mm for brick superstructure).

2. Slab thickness : 25mm with single layer 38mm chicken mesh
reinforcement.

3. Straight section from outer edge of end of 960 mm radius to
tangent point with 560 mm radius.

SPIRAL GEOMETRY
(radial dimensions to INNER wall of superstructure)

PLYWOOD TEMPLATE
(for first course of brick superstructure)

UNDP Inter-regional Project INT/81/O47

ZIMBABWEAN VIP LATRINES

FERROCEMENT AND BRICK SPIRAL LATRINES:
Spiral Geometry, Template and
Roof Slab Details

Dimensions in mm   Drg. No. ZVIP/03

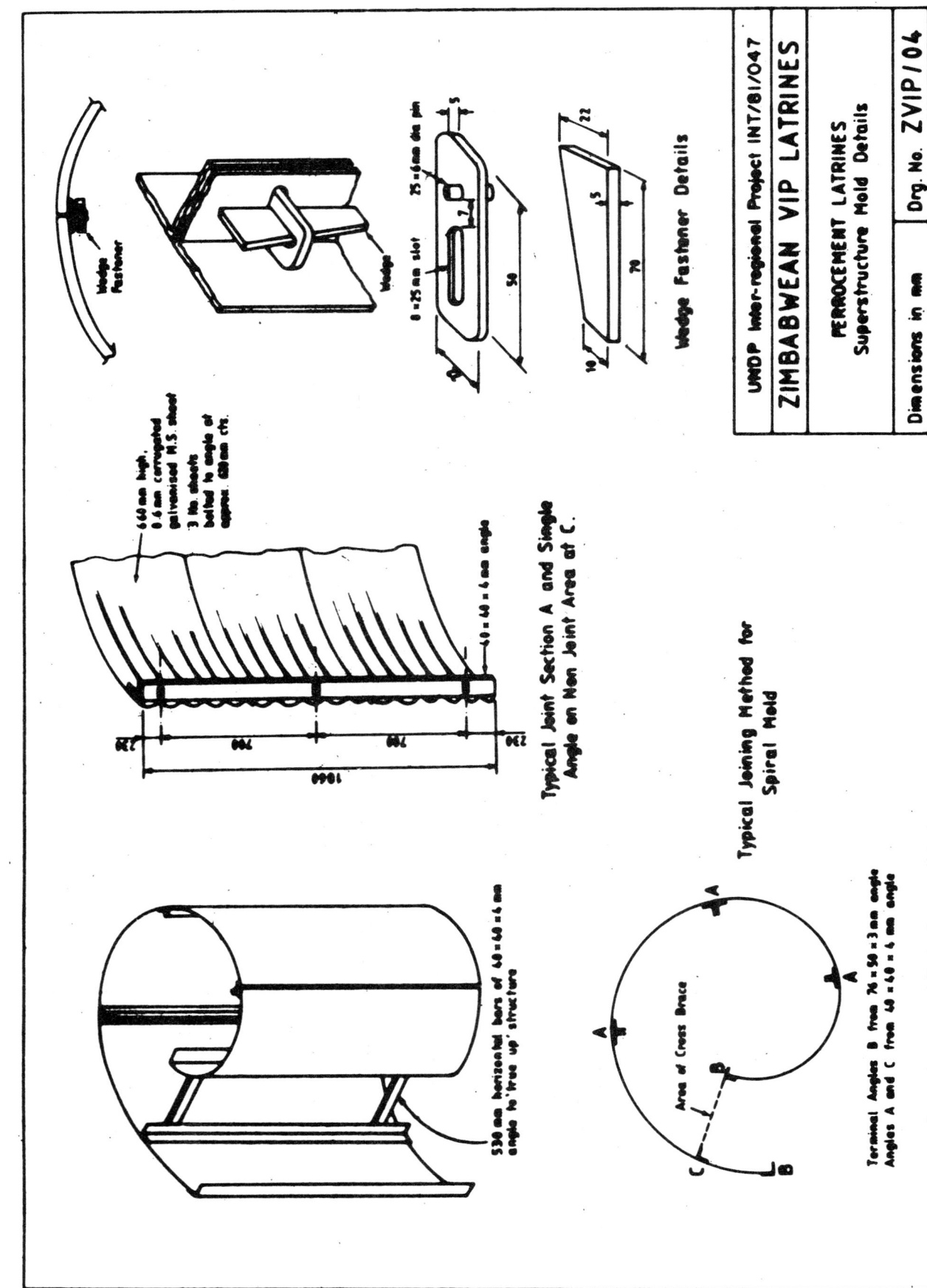

UNDP Inter-regional Project INT/81/047

ZIMBABWEAN VIP LATRINES

FERROCEMENT LATRINES
Superstructure Mold Details

Dimensions in mm | Drg. No. ZVIP/04

Wedge Fastener Details

Typical Joint Section A and Single Angle on Non Joint Area at C.

Typical Joining Method for Spiral Mold

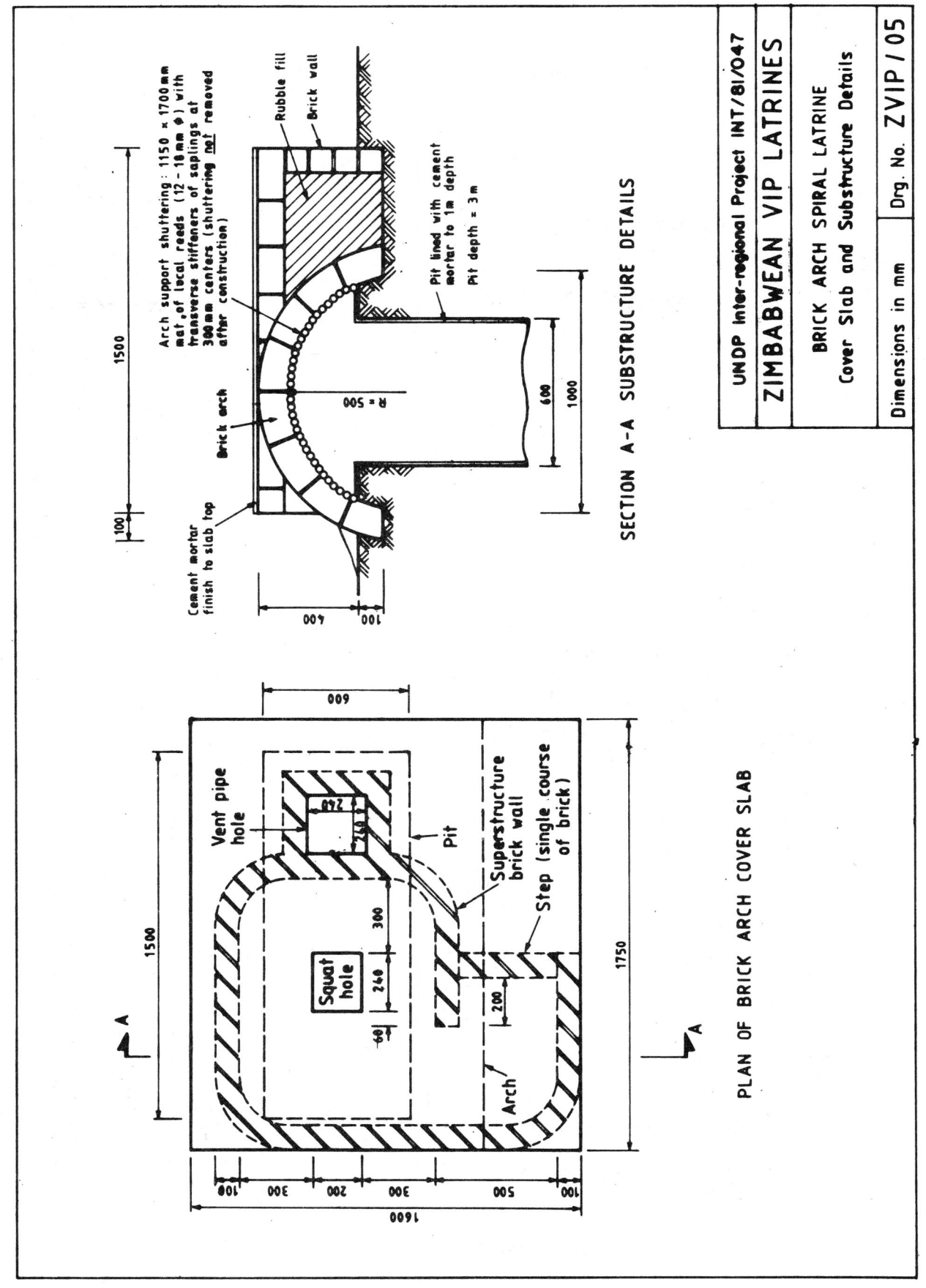

SECTION A-A SUBSTRUCTURE DETAILS

Arch support shuttering: 1150 x 1700 mm mat, of local reeds (12 - 18 mm ⌀) with transverse stiffeners of saplings at 300 mm centers (shuttering not removed after construction)

Rubble fill

Brick wall

Brick arch

Cement mortar finish to slab top

Pit lined with cement mortar to 1m depth

Pit depth = 3 m

R = 500

1500

100

1000

600

400

100

PLAN OF BRICK ARCH COVER SLAB

Vent pipe hole

Squat hole

Pit

Superstructure brick wall

Step (single course of brick)

Arch

600

1500

1750

1600

100 300 200 300 500 100

240

240

300

240

60

200

A

A

UNDP Inter-regional Project INT/81/047

ZIMBABWEAN VIP LATRINES

BRICK ARCH SPIRAL LATRINE
Cover Slab and Substructure Details

Dimensions in mm   Drg. No. ZVIP / 05

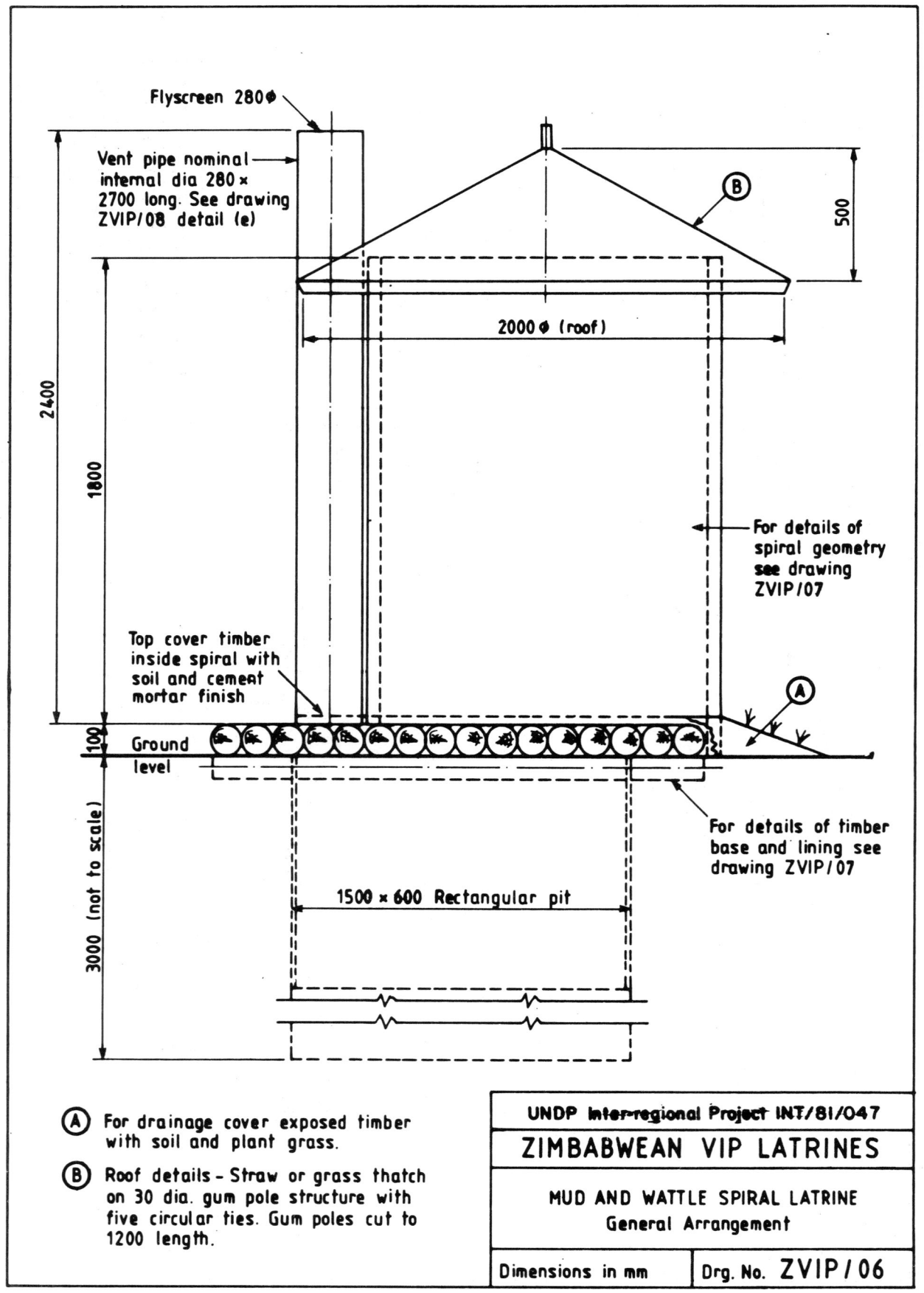

Flyscreen 280⌀

Vent pipe nominal internal dia 280 × 2700 long. See drawing ZVIP/08 detail (e)

500

2400

1800

2000 ⌀ (roof)

Ⓑ

For details of spiral geometry see drawing ZVIP/07

Top cover timber inside spiral with soil and cement mortar finish

Ground level

100

Ⓐ

For details of timber base and lining see drawing ZVIP/07

3000 (not to scale)

1500 × 600 Rectangular pit

Ⓐ For drainage cover exposed timber with soil and plant grass.

Ⓑ Roof details – Straw or grass thatch on 30 dia. gum pole structure with five circular ties. Gum poles cut to 1200 length.

UNDP Inter-regional Project INT/81/047
ZIMBABWEAN VIP LATRINES
MUD AND WATTLE SPIRAL LATRINE General Arrangement

Dimensions in mm	Drg. No. ZVIP/06

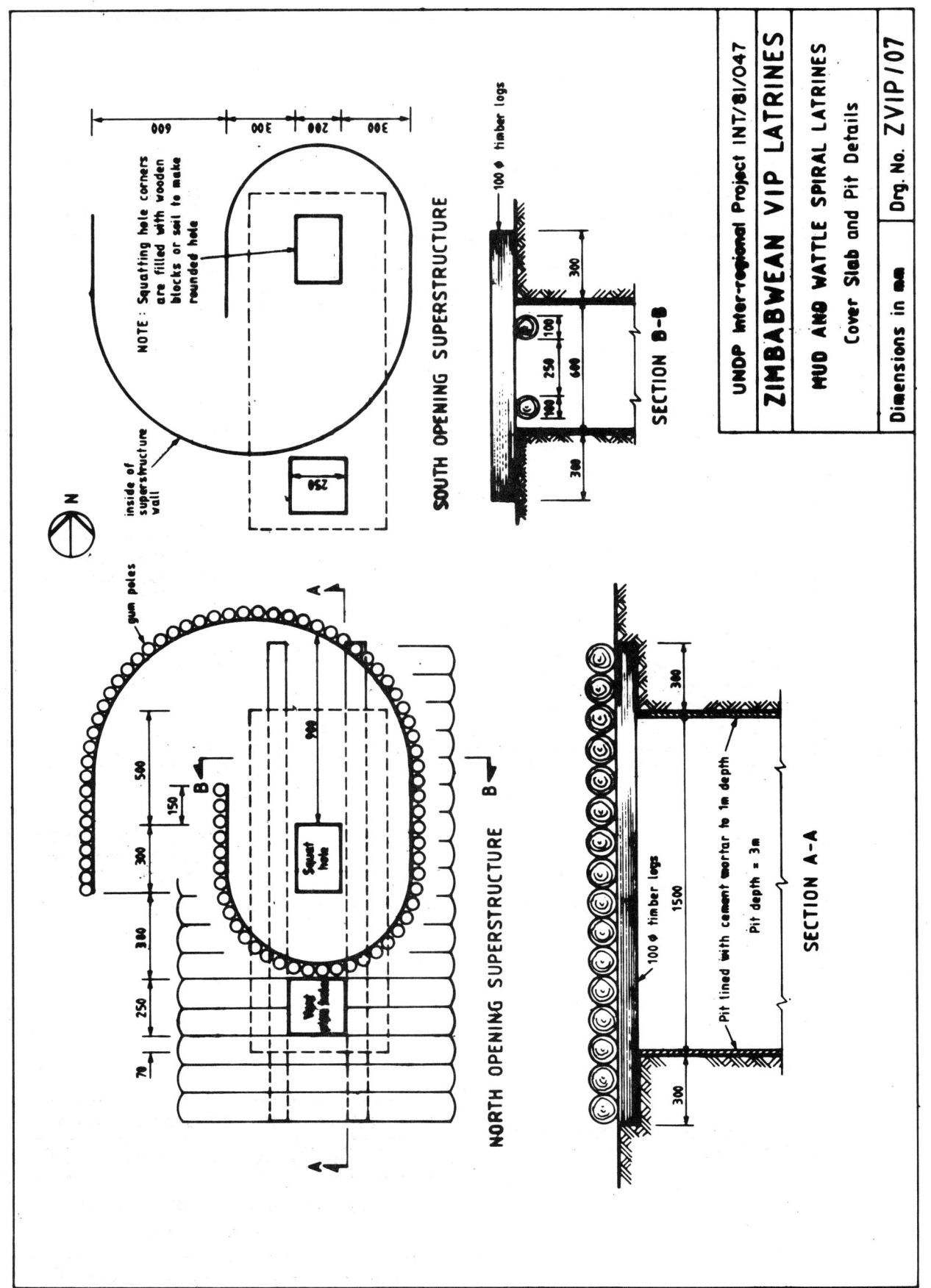

NOTE : Squatting hole corners are filled with wooden blocks or soil to make rounded hole

SOUTH OPENING SUPERSTRUCTURE

inside of superstructure wall

600
300 200 300

100 ⌀ timber logs

SECTION B-B

300

600 250 100
100

300

gum poles

NORTH OPENING SUPERSTRUCTURE

500
150
300
300
250
70

Squat hole

B

A

100 ⌀ timber logs

1500

Pit lined with cement mortar to 1m depth

Pit depth = 3m

300

300

SECTION A-A

UNDP Inter-regional Project INT/81/047

ZIMBABWEAN VIP LATRINES

MUD AND WATTLE SPIRAL LATRINES
Cover Slab and Pit Details

Dimensions in mm | Drg. No. ZVIP/07

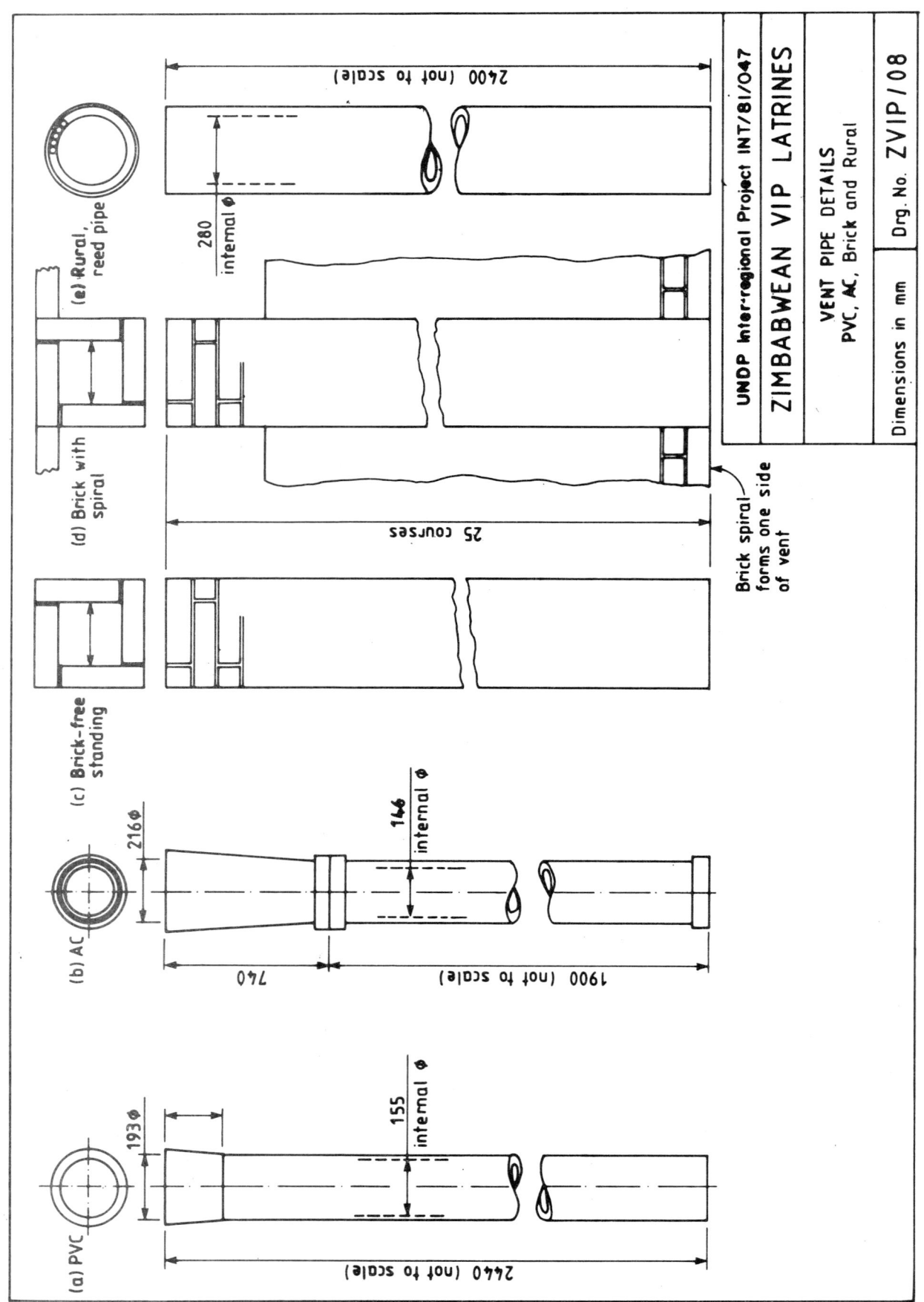

UNDP Inter-regional Project INT/81/O47

ZIMBABWEAN VIP LATRINES

VENT PIPE DETAILS
PVC, AC, Brick and Rural

Drg. No. ZVIP/08

Dimensions in mm

(a) PVC

2440 (not to scale)

193 ⌀

155 internal ⌀

(b) AC

1900 (not to scale)

740

216⌀

146 internal ⌀

(c) Brick-free standing

(d) Brick with spiral

25 courses

Brick spiral forms one side of vent

(e) Rural, reed pipe

2400 (not to scale)

280 internal ⌀

SCHEDULE OF MATERIALS

1.   **Ferrocement spiral VIP latrine**

Item No.	Description	Qty a/	Unit	Rate b/	Amount b/
01	Cement	5	50 kg	3.30	16.50
02	River sand c/	2/3	m^3	11.60	7.60
03	Pit sand c/	1/3	m^3	7.08	2.36
04	Aggregate (18 mm dia)	1/5	m^3	12.50	2.50
05	Steel bar (6 mm dia)	25	m	0.09	2.25
06	38 mm Chicken wire (1.8 m wide)	6	m	1.25	7.50
07	Bricks d/	50	No.	0.06	3.00
08	Vent pipe e/	1	No.	16.06	16.06
09	Wire (8 swg)	15	m	0.70/kg	1.05
10	Bitumastic paint (black)	1	liter	6.30/5 1	1.26
11	Superstructure mold f/	1/50	No.	126.30	2.53

TOTAL   Z$ 62.61

a/   Quantities for household unit (pit diameter = 1.5 m).

b/   Prices in April 1981 Zimbabwean dollars (Z$ 1 = US$ 1.52).

c/   Or builder's sand (1 m^3 required).

d/   If the superstructure is built in brickwork, a further 400 bricks are required (20 courses of 20 bricks); if the vent pipe is made in brickwork, an additional 150 bricks are needed (25 courses of 6 bricks).

e/   Asbestos cement vent pipe (see text, paragraph 21). The standard PVC vent pipe costs $ 16.50 with a 3.4 mm wall thickness and $ 13.50 with a 2.6 mm wall thickness.

f/   The mold can be used for 50 latrines.

2.   **Rural mud and wattle (or thatch) VIP latrine**

Item No.	Description a/	Qty	Unit	Rate b/	Amount b/
01	Cement c/	1	50 kg	3.30	3.30
02	55 m galvanised wire (18 swg)	0.5	kg	106	0.53
03	Fly screen (300 mm x 300 mm)	1	No.	0.13	0.13
04	20 nails (100 mm)	0.25	kg	1.02	0.25
05	20 nails (150 mm)	0.5	kg	1.02	0.51
06	Bitumastic paint (black)	0.5	liter	1.26	0.63

TOTAL   Z$ 5.35

a/   Items listed are those not freely available in rural areas. Each household is assumed to be able to provide local building materials (timber, termite-hill soil, thatch, reeds for vent pipe etc).

b/   Prices in April 1981 Zimbabwean dollars (Z$ 1 = US$ 1.52).

c/   If the superstructure is made entirely from thatch, only 25 kg of cement are required.

3.   **Mix Details**

For convenient reference details of the mixes of the cement mortar used for various purposes in VIP latrine construction are given below. In Zimbabwe, large galvanized buckets are used to measure out the required quantities of cement and sand; one bucket (bkt) contains approximately 25 kg of cement. Mix details are as follows:

(a) Pit collar and lining down to 1 m: 1 bkt cement, 5 bkts sand.

(b) Pit lining from 1 m down to 3 m: 1 bkt cement, 8 bkts sand.

(c) Cover slab: 2 bkts cement, 4 bkts river sand, 8 bkts 18 mm aggregate.

(d) Superstructure: for both ferrocement and brick designs, 3 bkts cement, 8 bkts river sand, 7 bkts pit sand.

(e) Roof: 1 bkt cement, 3 bkts river sand.

(f) Cover slab benching: 1 bkt cement, 3 bkts river sand.

4.   **Labor Schedule**

In Zimbabwe one builder and two laborers generally work to the following work schedule for the construction of one ferrocement VIP latrine (excluding excavation):

Day 1   Cast slab and roof; lay brick collar; plaster collar and pit wall.

(Day 2   Other work).

Day 3   Erect superstructure mold and cover with chicken wire; move slab on to collar; plaster mold.

(Day 4   Other work).

Day 5   Remove mold; place roof and vent pipes in position; cement benching to cover slab; touch up.

The schedule for the brick design (with concrete cover slab) is similar except that the work specified above for day 5 can be done on day 4. For the rural mud and wattle design, the schedule for 3 laborers is (excluding excavation) is as follows:

Day 1   Place logs and erect timber superstructure.

Day 2   Application of anthill soil to slab and superstructure.

Day 3   Make roof and vent pipe; plaster one half circumference of vent pipe; second application of soil to slab and superstructure.

Day 4   Fit roof and vent pipe; plaster other half circumference of vent pipe and cover slab.

Day 5   Paint cover slab with black bitumastic paint; apply soil over exposed logs and plant with grass.

MOSQUITO CONTROL DATA

Recent experiments in Zimbabwe have shown that mosquito breeding in wet pits can be substantially reduced by the addition of 1 kg of 4-6 mm diameter expanded polystyrene balls, thus confirming the work of Reiter [1]. The results obtained in Zimbabwe are as follows:

	Number of mosquitoes trapped during		
	Days 1-7 [a]	Days 8-14	Days 15-21
CONTROL PITS (no polystyrene balls)			
Pit 1	2549	2747	1438
Pit 2	1549	2345	1377
EXPERIMENTAL PITS (with polystyrene balls)			
Pit 3	1283	551	64
Pit 4	2162	583	66

[a]   Polystyrene balls were added to the experimental pits at the start of day 8.

[1]   P. Reiter, "Expanded polystyrene balls: an idea for mosquito control", Annals of Tropical Medicine and Parasitology, 72(6), 595-596, 1978.

BIBLIOGRAPHY

Publications by the Blair Research Laboratory

1.   P. R. Morgan (1977). The pit latrine - revived. Central African Journal of Medicine, 23 (1), 1-4.

2.   P. R. Morgan and V. de V. Clarke (1978). Specialized developments of pit latrines. In: Sanitation in Developing Countries (Ed. A. Pacey), pp. 100-104. John Wiley: Chichester.

3.   P. R. Morgan (1979). A ventilated pit privy. Appropriate Technology, 6 (3), 10-11.

World Bank Series "Appropriate Technology for Water Supply and Sanitation"

(Vol 1)   Technical and Economic Options, by John M. Kalbermatten, DeAnne S. Julius and Charles G. Gunnerson (a condensation of Appropriate Sanitation Alternatives: A Technical and Economic Appraisal, The Johns Hopkins University Press: Baltimore and London, 1982).

(Vol 1a) A Summary of Technical and Economic Options, by John M. Kalbermatten, DeAnne S. Julius and Charles G. Gunnerson.

(Vol 1b) Sanitation Alternatives for Low-Income Communities: A Brief Introduction, by D. Duncan Mara.

(Vol 2)   A Planner's Guide, by John M. Kalbermatten, DeAnne S. Julius, Charles G. Gunnerson and D. Duncan Mara (a condensation of Appropriate Sanitation Alternatives: A Planning and Design Manual, The Johns Hopkins University Press: Baltimore and London, 1982).

(Vol 3)   Health Aspects of Excreta and Sullage Management - A State-of-the-Art Review, by Richard G. Feachem, David J. Bradley, Hemda Garelick and D. Duncan Mara (a condensation of Sanitation and Disease: Health Aspects of Excreta and Wastewater Management, forthcoming. John Wiley & Sons, Ltd.: Sussex, England).

(Vol 4)   Low-Cost Technology Options for Sanitation - A State-of-the-Art Review and Annotated Bibliography, by Witold Rybczynski, Chongrak Polprasert and Michael McGarry (available, as a joint publication, from the International Development Research Centre, Ottawa, Ontario, Canada).

(Vol 5)   Sociocultural Aspects of Water Supply and Excreta Disposal, by Mary Elmendorf and Patricia Buckles.

(Vol 6)   Country Studies in Sanitation Alternatives, by Richard A. Kuhlthau (ed.) (forthcoming).

(Vol 7)   Alternative Sanitation Technologies for Urban Areas in Africa, by Richard G. Feachem, D. Duncan Mara and Kenneth O. Iwugo (forthcoming).

(Vol 8)   Seven Case Studies of Rural and Urban Fringe Areas in Latin America, by Mary Elmendorf (ed.) (forthcoming).

(Vol 9)   Design of Low-Cost Water Distribution Systems, Section 1 by Donald T. Lauria, Peter J. Kolsky and Richard N. Middleton; Section 2 by Keith Demke and Donald T. Lauria; and Section 3 by Paul V. Hebert (forthcoming).

(Vol 10) Night-soil Composting, by H. I. Shuval, Charles G. Gunnerson, and DeAnne S. Julius.

(Vol 11) A Sanitation Field Manual, by John M. Kalbermatten, DeAnne S. Julius, Charles G. Gunnerson, and D. Duncan Mara.

(Vol 12) Low-Cost Water Distribution - A Field Manual, by Charles Spangler.